The Genius Machine™ and the Theory of Everything

By
David Walls-Kaufman

Once Again, for Kimble

Ecclesiastes 9:15 "The wisdom of the poor man is despised."

"I give solutions here to the experts who have no solutions, but they ignore what I say because they say I am no expert."

—Dr. David Walls-Kaufman

Paperback

ISBN: 978-1-966606-22-2

Hardcover

ISBN: 978-1-966606-23-9

Part I:
Existence—Life Ascends to the Genius Machine™

Article No. 1:
How Can We be Truly Happy and Make Sense of Our Life Without Knowing Why We Exist?

What is the meaning of life? What is the Theory of Everything? The Theory of Everything is the attempt to find and describe the centerpoint of existence, and human existence, around which *everything* forms a circle with a single point of function, gravity or orbit, like our Earth around its gravitational center, and our solar system around the sun. But not just physical matter. *Everything.* Everyone has thought about the meaning of life. Many have said it is love. But how can it be "love" with so much strife? The answer is that we human beings do not yet realize that we live inside the hardened shell of an existence that is a machine *forcing* us through the consequences of our actions to realize what we are best at, and how we would live the best life. This book describes that machine. The great machine functions around the purpose of both our thriving and our highest creativity to amuse and improve ourselves and to learn about the world around us. The first important lesson is to understand that our existence is this machine with a hardened operating system that *will not* change because it already makes us the perfect offer. The more we are too stupid to see the perfect offer the more we will suffer. We will continue to suffer until we see what this perfect offer is, and that it is not possible to come up with an alternative that is better than the one already being offered, and that is seemingly eternally offered. We can never make the most of our lives without understanding the function of the machine and that it exists only to offer us the most splendid holistic existence we, as mortals, cannot possibly imagine an existence greater than. It is as if the universe is telling us, "Wake up, friend—everything you have the capacity to want, and are entitled to want, is already waiting for you."

This is a matter of surviving and thriving. It is a matter of health. If you are expert enough in the subject of health—the nature of the hardened shell of the operating system becomes clear. If you aren't an expert in health then you have no idea what you are looking at. The world is a cruel jumble. But if you know that stress is the cause of disease, then you have the clue to begin to make order out of the mayhem. Nothing is more daring than to propose an observation on the function, orbit and meaning of all that is. It becomes easier the more you look at the world that there just is no other way to assemble all the pieces, as you will see for yourself. The gift revolves around thriving. Thriving is measurable. It is different from anything less. A biological determinism frames thriving because health and thriving are something earned not given. We know that we cannot do anything we want and earn them. Even if a mortal has thriving they still have biorhythmic ups and downs. Yet their lives are still special in every way it is conceivable to measure. I could get credentials in sociology, psychology, medicine, public health at a university, and then ask for a grant of $10M to fund research for 10,000 different measurements of human thriving. But we already know the things I am talking about. What's worse, we now know that actors in medicine, medical insurance, Pharma and Big Food, universities, media and government have colluded to keep us ignorant on simple holistic matters of health, and profited stupendously in doing so. As it is, with this atrocious situation, this book is speculation, a piece of philosophy, not science. But the sciences already fully corroborate "holistic" ideas, just as your own common sense tells you very well what is holistic (thoroughly good), and what falls short. The more you know, the more you know that what is holistic is better than what is artificial, when it comes to things we put in our body. As in the old saying "an ounce of prevention is worth a pound of cure." We would rather never take a drug, or a surgery, or an implant, or any offering from Pharma, if we did not have to. If you know enough about health, medicine, disease, and thriving, you know that what is holistic and true in the end boils down only to how we live and think, and whether or not what we do and think matches up with what is already here, waiting for us, in the universe. Every answer we desire comes down to how we live and think. More than that, the right answer always boils down to a choice, and whether or not the choice we make is congruent with this universe, this existence, this miraculous machine that we live in, that we were born into.

By the end of this book, you will find for yourself that the explanation and machination of existence and human existence is so beautiful, obvious and deductible that they cannot be argued against. *That* is the ultimate proof of why you should quit fighting your life.

Article No. 2:
Why is there *Anything* Rather than Nothing?

The beginning of any discussion of the Theory of Everything or Unified Theory must start with Leibniz and Spinoza, who both asked, "Why is there something rather than nothing?" Indeed, why is there *anything* rather than nothing? Victor Frankl, in his classic *Man's Search for Meaning*, said that he never found the meaning of life, but only meaning for people specific to different situations in their lives. Here it is: The secret, meaning, reason, purpose and methodology of human existence is first to simply exist, so that we stick around long enough for us to eventually discover that the offer made to us by existence, by God, is so exquisitely exceptional that, given all parameters, it *cannot* be made any better or any more interesting than it is. We do not grasp this yet because we see so much suffering, pain, misery, poverty and cruelty. Or because we want so many different things in our lives that we will probably never get. Most of this exists (except for accident) because we do not yet see the perfect rules that confine our existence in order to guide us. Those rules are like walls made of poison. Like stress. The more you bump into them because you don't know what they are—you poison yourself. If you knew what the rules were, and that the barriers were made of poison, and that they were *never* going to change because they were so perfectly good for you and all the rest of humankind, you would soon lose the desire to hurt others and lean into them. Just like not sticking your hand in a fire. Except in this case the stress-change probably alters the quality of life of everyone else in the human set. Just like cells in the same body. Or water molecules in the same pool. Eventually, pissing in the pool changes the whole pool. But there is a lag time while urine works its way around. However, in a quantum way the entire pool is changed as quickly as it's pissed in.

Because the answer to everything is the most complex subject known to date, let's put all the cards on the table and see what you want to dispute: The human race (indeed *all* of life wherever it is created) is always on its way to discover that it is a Genius Machine™. All of our doubt, confusion, history, and man's inhumanity to man is the spitting, resistance, vibration, coughing and uncertainty of any one-stroke lawnmower first kicking into life to begin to feel out its function and mechanics to eventually learn what its highest, most efficient and glorious purpose is. Nothing delights us, entertains us, improves us, or interests us more than our own genius. Our genius in *anything* so long as (curiously) our genius is not itself used to interfere with genius. We are on our way to discover that the human race is not made of many tribes, but instead made of essentially *one* tribe made of many charming sub-tribes, whose process is then made much more efficient, and our own great variety involutes to further stimulate the genius machine far more than sameness ever would. How do we find all of this out—unless the greatest lesson of all time is just supposed to be *given* to us like a piece of trash that we would then disrespect with all of our typical disrespect?

Let me further translate the invisible rules of structural force that are inarguable because they refract back into *your* best life. They say: Harm to none, thriving for all, in order to create the optimal platform for genius and the most satisfying, rewarding, fulfilling, safe, holistic, risk-free, healthiest, natural, unartificial, fun, joyful, the most *interesting* life experience a mortal can achieve, or even *imagine* as a reality. Violate this formula in any way and you take on resistance, stress and risk. *Someone* will wait outside your hut. This stress is not only the cause of all disease, it is the cause of all our problems. Stress implies that there is better and worse. Better and worse implies that there might be an ideal. And if there is an ideal then, as Harrison points out, there is only one. If this is true, then the argument is over. The data will never fail the lesson. To challenge this life, or even to pretend this perfect offer is not made, is to stress ourselves. What we suffer only happens to us so that we do minimal damage to ourselves until we wake up to the facts about the best way to spend our time here. The universe makes it in *your own interest* to find the answer as quickly as possible. This best *potential* waiting in the operating system of existence appears to be the only thing in the universe that does not change.

There you are—the destiny and evolution of all life summarized in one page. To argue out every possible protest will take some time, but that is it. First, to find this lesson, we must exist. Second, we must be born at the opening of a tunnel of hardships that is as long as we choose to make it, designed to make us learn and appreciate the unique gift that has been given the moment we come into existence. Seeking to avoid pain, disease, stupidity, risk, stress, resistance and conflict is to realize that *all* of these problems have only one source. Likewise, in mirror image, the *only* escape from all of these problems is—the purest tuning of human survival, thriving and the Genius Machine. Three things we are most interested in, that we like and love most, are our survival, thriving and genius. These interests fine-tune the machine. Everything we dislike and loathe, and that ruins us, progressively interferes with the machine. We have no choice. The model is already in our DNA, in our psyche, in our consciousness. The structure is already there so much so that you cannot argue with it. The progressivity of this phenomenon looks exactly like elastic integrity or elastic recall—the farther we push or stray from the pure center, the more strain, stress, resistance and problems we encounter. These problems are our guides. The model gives us the very definition of sanity and insanity. To oppose the genius machine, the more you oppose it, the more you are insane and sociopathic.

At this point, some people argue that the perfect moral model is blown up by the fact that they can't get all they want—like a mansion. But they only want a mansion because genius made it. Without genius, they wouldn't even have the aspiration. Let's explain our own disappointment in not getting all we want. The answer is that there is a world of people out there waiting for you to raise their quality of life. If you do it enough, they will reward you in a free exchange enough that you can get the car or house that you may desire. This meritocracy is innate and cannot be made any better, any more fair, more decent, more humane, than it is. Anything else is stealing. Nothing is free. To get more of what you may want, you must elevate others. And the only way to truly elevate others is to do so holistically, i.e., in ways that *never* hurt their thriving and that *never* impede the Genius Machine. Any deviation trembles the machine. Any deviation pisses in the pool.

Article No. 3: Let Me Try to Kill You—And You Let Me.

Take for example if I try to kill you, or a family member or friend. Or if I try to take what is yours. You fight back and resist in proportion to how much I'm trying to hurt you. *Why?* Because you don't want to lose either your life, or the life and presence and support of someone you like or love and who supports you back. And you do not wish to lose what you have earned "honestly" by helping your tribe, and that pleases you and makes your life more secure. *Why?* Says who? What difference does it make that you are here and more comfortable, or not? Why do our survival reflexes make us fight back and want to *be* here? We simply *accept* this Maslow's hierarchy of needs as if we matter, when, in truth, just like existence—this moral hierarchy doesn't have to be here at all! But it is. Despite ourselves. And it is aimed at the discovery of the genius machine. It is as if something wants us here to stick around long enough for the greatest discovery ever.

The more I threaten you and yours has an elastic property. The more I do it, as long as you *know* I'm doing it, the more you resist me. To you, there is no call for my threatening you. You don't deserve to be treated this way. *Why not?* We didn't make ourselves. In fact, we can all say—because history has taught us this lesson—not one of us is here to have their lives made less than what they could be. This is a marvel of commonality! *How did this remarkable set of survival reflexes and expectations come to be?* Suddenly, we have a universal definition of innocence and guilt. He who helps the tribe is innocent, and he who hurts the tribe is guilty. How did this ideal moral universality roll out of randomness, chance and chaos?

All tribes define innocence and guilt the same way: How much is the tribe member distracting or hurting other people in the tribe from helping with the tribe's survival and thriving in order to be in a better position for what comes next. And when it comes to "what comes next" what is always the best thing that can come along?

Genius in anything.

The elastic predictability around what is best, most entertaining, most rewarding and interesting for us is as certain as a mathematical formula. Our behavior shows a hardened symmetry and circularity, an envelope, a sphere, the more we zoom out. What you like least is obvious and predetermined, and what you like most is obvious and predetermined. And there seems to be no way to make it any better than it is.

Article No. 4: Are We Here to Bear Witness to the Exquisite?

The idea becomes apparent when we start talking and exploring along these lines that truth starts with two rules. The first rule is First, Do No Harm. The second rule of truth is that the spherical shape of truth is so big and powerful that it can be readily found in instinct, data, history, tribal law, consequences and results that it can be readily discovered and outlined with common sense and an open mind, without the hate and violence that bad influences try to persuade us with.

We are here to exist, survive and bear witness to the miracle of creation and existence. Everything about us forms a spherical fortress around doing this one thing. The highest potential of creation is to evolve living things to the point of sophistication where they suddenly dawn onto the realization that life and existence untiringly bring the perfect offer of the perfect experience. And therefore it is wonderful that there is something rather than nothing. We are created to be introduced to life and existence, to experience them, and to explore, affirm and enjoy the *real* fact that existence is this whole truth. Our lives are linear and predictable around this point, this orbital centerpoint, except for accident (and genius). Accidents are inevitable because free will and mortality are essential in the making of a thing (us) with problems and limitations that lead us toward what is at once most safe and interesting in a positive feedback loop. Free will makes it inevitable that things will bump into each other, the basis of accidents. Without free will the grand lesson of existence means nothing. The invisible rules of the universe, that science calls "fundamental physical constructs", or what we might better think of as *structural forces* of the universe, frame both this constant offer of an ideal, and also the ideal mechanism for finding the offer as quickly as possible, since the offer *cannot* be given for free and remain the most important realization there is. All pain makes the lesson stick, makes us seek solutions.

That we may not *choose* the perfect lesson, or reject it, or that we are unaware of the offer, does not break the perfect offer or the perfect mechanism. Nor is it true that genius is the only reward, motivation or interest for anything, but that genius sits at the top with anything less priced in descending order down to the last penny. And any artificial enhancements or substitutes, after the hormetic buffer, transform into a stress that undermines our structure, chemistry, psyche and function on a two-way street of bounded instability, framed by the emergent self-order and spontaneous order of complexity science and systems theory observed by Driesch, Powers, Bertalanffy and others first discussed as early as 1803. All of it exists *only* to lead to the most wonderful revelation that mortals can imagine. The totality in coherence and organization appears to exhibit modular integrity, tensegrity, auxetic integrity, superrationality, equifinality, isotropy and monism. Whether this is a phenomenon of chance or not matters little in that the moral and motivational grid is inarguable since it has no place to evolve to. And it does strain credibility to argue that anything so super-humanly designed, that would appear to be morally permanent, could be the product of randomness, chance and luck.

Chance is rolling the dice and getting four sevens in a row. That does not happen in our universe, at any level. Reality would collapse if it did like a rickety ladder. No. In our reality, every constructive iota holds serve so that everything built on top keeps shifting toward realization.

No one can or will argue the point because there is simply nothing more that mortals can ask. Not without asking for superpowers that would only give them the power to abuse others. Around this centerpoint is framed the human psyche, instinct, biochemistry, structure and function, even social structure. The more I talk about it the more it reveals itself to you as being *obviously* the only basis of everything, at least in human scale. This moral structure is the basis of stress and benefit. Collectively, stress and benefit form a "sleeve" of bounded instability in which the entirety of a human being is moved eventually either toward holistic thriving or the opposite *only* in proportion to congruence or divergence from the ideal. Stress or benefit, congruence or incongruence, are the sleeve that either braids or unbraids the rope. The rope of psyche, function, chemistry and structure. Our body re-braids our rope as best it can after any damage to keep us on our course. Our opinions and personal tastes on what stress and benefit are make little difference. What are your opinions on black mold hidden in your walls? Or falling from 50 feet? Or strychnine? The ideal makes "majorities" and moral relativism obsolete. This incredible humanistic reality in the ideal of the human genius machine becomes more self-evident the more we examine all facts and evidence.

Every piece or wave-form of the "inanimate" world aligns in layers, blocks or braids to build this platform for life to begin evolving toward the greatest lesson that can be realized. And so, of course, it is always better that there is something to see it and enjoy it rather than nothing. Something so morally perfect and beautiful *should* be appreciated.

This is the neurophysiological setpoint around which our entire life orbits, including economics and value down the slope from genius to the last penny. We have no real choice in the matter. Price is determined by what interests us. Nothing is more interesting to us than survival, thriving and genius in anything. *Something* made us like what we like—and as long as we do not hurt anyone else and disrupt the genius machine, it's okay. It is astonishing that even these liberal parameters cannot be made any more liberal either. It is as if something *likes* us and *wants* us to be here, and curious and active. The operating system works linearly and predictably around this centerpoint process, and all of us have a good instinctive sense of what the specific morals and constructs are that create the platform for optimal creativity. Good data resolves disputes; except for the insane. Data says return to basics. Keep it simple. Mind your own thriving and do not hurt others with it. This construct is not only the basis of reality, it is the fabric of the phenomena of information, logic, knowledge, neural nets, common sense, confidence and trust, instinct, learning, economics, justice, right and wrong, innocence and guilt, all results and all facts.

We *only* have problems because we fail to comprehensively understand "First, Do No Harm," the 1835 motto of the American Homeopathic Association, and Thomas Jefferson's "All men are created equal." This rule is, apparently, not fooling around. This means *no harm* in an exacting, holistic way that challenges you to think outside your short term self-interest. You must *actually* do not harm. As defined by a power that can even see inside your own mind, probably, since we can do ourselves psychological harm by thinking the wrong way. The rules appear able to detect your most secret wish to upset the uncanny rules that, incredibly, appear to protect and promote the holistic interest of every last person that is here. It seems that, only if we *voluntarily* obey these unwritten rules then the perfect life unfolds for us from inside out, just like all other nuances of health and thriving, spontaneously, irrevocably. Well-being and fulfillment appear to spring like a guarantee in trade for you doing what you are supposed to, for doing what is best for you anyway, and nothing else gives you true vibrancy. Any niggling miss or overreach *bothers* you—as we can see in all the fake, half-conscious humanistic frauds all around us who slowly go crazy. We see their faculties degrade the more they lie and inveigle against The Good.

This is the only conclusion that can be made given what is now known about history, politics, health and physiology. Yes, there is the horrific cruelty of the Communists, German National Socialists, and radical Islamists. But why are we averse to risk? Nothing is more risky and costly than tribal conflict and war. Why are some of us early adopters who are never directly threatened yet choose to be our brother's keeper and fight for a better world? And why do we have an instinct to raise our quality of life? And why do we have a righting-reflex to fix our mistakes? And why are all these problems solved only by a timeless ideal that asks us only to be better than we were first born, as animals. Born into a world of dog-eat- dog, and kill or be killed. A vicious world that is always there for you if you choose it. A world that is itself inevitable in order to learn; in order for any of this beauty to last and really mean anything, or stand for anything. We take this savagery to be an inevitability in the human condition. But the Epiphany at Sinai in 1313 B.C. *told* us that improving our world depended on how we perceive one another and how we treat one another. All of our reality changes with merely a choice. Really, our only argument to resist this deal is for us to point out how we are not sociologically used to this clear choice yet, how the finality of the choice is rather shocking.

It appears that all of our answers are *only* found in a perfect held in place as a constant by every molecule of everything. Or every part of whatever makes everything.

The only way that existence appears not to be perfect in this mechanism and in this offer is the way in which so many of us suffer and die prematurely, from disease for instance, before we open our eyes. We will address this later. But our innate morals point out that it is not enough to sit and simply gawk at creation. Gawking at creation is part of the joy, but the next part is even better—to fully participate in creation by being a creator yourself as best you are able. That which is created is created to create. Nothing will interest and fulfill you more.

Article No. 5: Nothing Exceeds Creation and Existence

The meaning, secret, purpose, reason and methodology of life and happiness begins with existence. If you do not exist you cannot be here to observe and care. If you exist then, at each moment, you must continue to exist, and exist well enough to function, and breed. To breed requires health. As has been said, the healthy man has many dreams; the sick man has only one dream. You must breed and have a lifespan longer than a shrimp unless you can figure out all of the magic of life and creation in a day, and then hand off your notes on the astounding patterns to the next generation so that they too can derive maximum benefit from what you have found in life's blueprint. Without health, nothing can be built on top. Missing health, you can be in such oozing, wrenching agony that you *cannot* focus on anything else. You were put here to thrive. All the tools are here for you to work and earn thriving.

At your holistic best—which is deceptively easy to achieve—you do not cheat a single other person of *their* thriving. But we are a creature born without parents who knew the lesson. Therefore, the indirect and direct consequence of *every* thought and action yields consequences that are our teachers. No part of us can hide. Existence is the police. Just like cells in the same body, or happy members of the same small tribe, or vibrant molecules in the same engine—there is a law of integration in any organ, system or individual. Only lag times appear to separate us. Eventually, there is only the illusion of separation. Every part of your holistic thriving costs others nothing. Most interestingly, at our best we each thrive the more we each thrive. We know this is true at least because of two laws: The law of Balance, and the law of interconnection. No part of any system, organ or person thrives or dies alone. The sum of their thriving or dying is always the sum of how their thoughts and actions have aligned or split. It would seem that consciousness is the plasma or scaffolding that builds the world. If something lies between consciousness and nothingness then this fact makes no functional difference. As soon as existence takes shape, it would appear to press everything born into an evolutionary flow toward realizing that which is always in *your own best interest* to know.

Go out there and get it, play with it, as hard and as much as you want—just don't hurt anybody else or their chances to get it and enjoy it. It is a meritocracy of genius: Raise one another's holistic quality of life to better serve genius. Any other way misses the near for the far or the far for the near. And the most important ingredient for finding our complete answers is the open mind: What is right, and how do you know when you have found it? (Because you do no harm).

Truth is NOT in the eye of the beholder nor morally relative. The Truth is a rather easy, discernible, physiologically provable, duplicable, constant, universal rallying point for Order that will crush any resistance or obstinacy or solve any disease or pain. Zoom out to three hundred miles, and we behave exactly like a machine. Miss the holistic ideal by *any amount* and we will be off the ideal by *that much. Anything that can go wrong will proportionately start to go wrong in bounded instability*—the unraveling of the rope. The hardest part is to try to predict exactly how monstrosity will manifest in any enormously complex living organ, system or individual. All we can be sure of is that, because of Balance and integration—every iota of benefit or stress is precisely accounted for. Anything else would not be fair. Any alternative would make revelation *impossible* to figure out.

At this point, which do we think—that the world is chaotic, or partly chaotic, or that it is perfectly humanistically ordered? We know all of this is right because we are trapped by the alternatives.

Part 2:
Is There Any Real Choice with an Ideal?

Article No. 6:
Two Parts of the Same Dynamic: Thriving and Genius

Notice that health is so essential to our existence that it is the entire focus of the Left-Right war that threatens to devour the West. The two great gladiators are Communism and western civilization. The Left, Communism, Socialism is at war with western civilization, and civilization in general (our status quo) because problems exist and the Left has a blind faith that wealth redistribution and free healthcare and generally making things easier for human beings is always a good idea. But "easier" is not always the answer in living systems. "Easy" may lead to mediocrity. What thriving requires cannot be fudged, ignored or deviated from, if it exists. The Left wants more social welfare to build up the complete mind, body, spiritual health of those who suffer and who have the least opportunity to make the most of themselves to explore fully-rounded citizenship to contribute their best to society. The Right says that, yes, social welfare is well and good—but how much welfare (which is highly addictive and is *not* thriving) do we give before free money becomes injurious to the citizen, and we pull resources away from genius? Helping people and helping genius are both immaculate arguments. But cheating any requirement of thriving will never help genius. The problem is we see thriving and genius as mutually exclusive. They *are* mutually exclusive, and they will *remain* mutually exclusive, if we fail to live and think holistically.

How do we pick between thriving and genius? We cannot,

because thriving and genius are equal parts of the same dynamic, optimal life. It makes no sense for one part of the genius machine to attack, cheat, sabotage or undermine another part of the genius machine. Again, the miracle of data, information and knowledge only exist to help us find the two-stroke engine of thriving and genius, and what each cannot help but mean to the other. The scientific evidence for this claim is already well on the way and probably finalized—but Pharma and Big Food paying so much money into Washington to corrupt research, publishing and advertising about health has made it impossible to prove and publicize the holistic data on a scale to unstick society. For these reasons this essay must remain a piece of theory, of philosophy, although I am sure the science will back it up—just as we are sure any sane person will not stand idly by while any insane person tries to kill them, steal their holistically earned property, or undermine their health, or join all people around one worthy goal. Once we have made our health linear and predictable, the only real question left is, "How do we best use all that thriving?" The answer should be inarguable and obvious. Existence appears to align itself in a protective circle-shell around the perfect. At some point enough puzzle pieces are in place to visualize the circumference of a flawless moral circle.

Article No 7: Human Life is Linear and Predictable Except for Accident and Genius

Thomas Szasz said the world of scholarship and academics is like a movie Western: He who "quick draws" the right statement, concept or language first is the winner. The unified theory is that our actions, motivations and level of interest are framed on one hand by death and suffering, and our desire to thrive, and then on the other hand to pursue our highest level of amusement, improvement and interest. Many theorists like Marcello Gleiser and Isaiah Berlin say that finding a single lynchpin for the meaning of life is impossible at least because science cannot reconcile science and the humanities, let alone make sense of human history and cultural diversity and the laws governing gravity versus cellular physiology. Berlin called this the Ionian Fallacy as his counterpoint to the monism of the Ionian Fantasy that was the ancient hope that the function and meaning of life *could* be distilled down to a single orbit. All of the "inanimate" world is the platform and art gallery, the "dead" platform, from which life springs to discover ultimately what is in its own best interest.

The quantum physics and astrophysics calculations outline the function of string theory. The spherical integrity of perfect human motivation takes its place in string theory, however that evolves, to capture the action of life. These interlocking blocks of perfect fixed reality take the string and form a circle. Nothing else would bring an ongoing series of breathtaking, reality-shaking surprises that will each be equivalent to our level of perception and understanding *at that time*. Every fact and law on the books acts to maximize this journey of surprise and appreciation. We are to be as equally pleased with this as God is pleased with us: For all of us to get the most moment-to-moment experiential epiphany along with God, Nature, Life, Existence, Creation, the Cosmos, the Universe, or Godness, or anything else we want to name or imagine that put us here and set us in motion. It is beyond comprehension to try to argue that this hardened, self-permanent super-human function is the child of chance. The stuff of chance, randomness and luck wink in and out of existence *constantly.* They are the perfect *opposite* of the universal consistency in the operating system as we know because *any* living system is always evolving to the moral epiphany where, for example, they realize that slavery and serfdom and diminishing the life of anyone is morally wrong. (They are *only* morally wrong because they do not coax the best out of us in the feeding and enjoyment of the genius machine.) Physicist Frederick Dyson said "It is as if the universe has been waiting for us." This is true. Our innate, quantum ethics *obligate* us to better treatment of one another *for the maximization of our own advantages.* The universe is waiting for us to discover the inherent perfect and quit hurting ourselves with our ignorance of the undeniable fact. This process requires that we *must* try to be better than we are born, and this *effort* to improve ourselves taps into a form of work and function that in of themselves rewards us with gains we will see by no other means. Astonishingly, the *only* medium by which this is possible is our own *choice* in how we treat ourselves and others. Does that sound about right to you? Miraculously, this defines stress and benefit in human scale. And, for any problems beyond human scale, the same benefits remain the best options to heal those as well. And if we get knocked off—life just keeps bubbling up toward the same ultra-beautiful conclusion. We each hold innate expectations for how we are to be treated, and they too conform to what we are best supposed to do.

Nothing is so audacious as to propose a Theory of Everything. A Theory of Everything is in fact the *answer* to everything. But as the point becomes more inarguable it becomes less risky. For instance, it holds the immediate answer to the cure for cancer. A sound Theory of Everything must tell you the cure for cancer as well as it tells you how to live your best life. What causes cancer and missing your best life? Stress. What is the cure for stress? Benefits equal to undo stress-damage already done. The ToE must give you every answer to everything—except the surprises yet to come. You don't get those until you're ready for them, it seems. For example, why does evil exist? Why do people not do the right thing? Why do people hurt me when I am a good person? In five thousand years of civilization no one has satisfactorily answered these questions. The answer is that even though the perfect offer is offered, it does not stop human stupidity or willful ignorance toward the perfect offer. It means that all sane people have a moral righting reflex not to like being wronged, and to search for what is right. The more people know what is decisively right then the more what is wrong is pinched into a space where it cannot hide. Like genius, this picture of creation is not really "ours"; this is another glimpse at another layer, another feature, of the structure *already in the design of existence.*

Physicists preoccupy themselves with the *how* on the function of life rather than the *why*, but at the right moment the scientific evidence and observable data tip the scale on *why* the human machine works exactly like a machine, and the *how* becomes only academically important. For example, *how* the forces that make the balls of a Newton's Rocker swing and click is a question and explanation as deep as we want to go in pursuit of an exact understanding of *how* consciousness and existence work. But *how* we exist, and *how* a Newton's Rocker yet again demonstrates Balance, becomes less important once we realize that these facts will never change the perfect *why* regarding our physiology and existence. Answers to *how* life works will never alter the perfect *why*. (We will never stand for it, eventually. Those who commit such a crime will be undone by their own conscience, eventually.) No sane person ever *wants* to change this perfect reality anyway. This humane balance *is* the structural force of the universe. And it *must* be this way because everything would fall apart if it didn't, and it would be horribly unjust if it did. The self-discipline the system asks for does not mean to be a monk. The following mathematical equation represents well enough how this two-part node of thriving and genius works that cannot sustain the tiniest edit:

$$\boldsymbol{\sigma} = 2\mu\boldsymbol{\varepsilon} + \lambda\ \mathrm{tr}(\boldsymbol{\varepsilon})\boldsymbol{I},$$

Here, sigma (σ or Σ) symbolizes how Lamé coefficients and Hooke's Law describe the modular integrity, auxetic and tensegrous tension, slope and run, curvelinear progression, elastic deformation, isotropy, monism, and the mathematics of elastic structures that actively hold all life systems in place, in any system, organ or being, including morality, tribal dynamics and the human psyche. Every biological system and organism shows evolutionary deference to this icon of modular integrity for the sake of an eventual higher holistic self-interest. Einstein's special relativity took the Pythagorean Theorem and Cartesian coordinates of 3D and added the 4D of space and time, into an X-Y-Z-*T* (Time) whole field that shows total organismic modular integrity around itself. What is proposed here is that the moral/ethical reason-for-existence is a 5D element in an X-Y-Z-T-*E* (Ethics) envelope for the final layer, moral push-pull closed system that probably is the end of the road of understanding *why*. The *how* is that the structural forces of consciousness dictate the outcome. This X-Y-Z-T-E formula shows modular auxetic integrity to itself. The math for Lamé coefficients and Hooke's Law are perfectly serviceable for showing action- reaction, but the hardened sphere of human morality could be symbolized by "C" representing the middle C note (tone) on a piano keyboard.

A tribe may be a primitive predatory tribe that sees its own greatest well-being coming from harvesting other tribes. But this is primitive and short-sighted: If any tribe, no matter how primitive, can be educated to become ingenius, or helpers of genius, then aren't they worth keeping around? If not, then humane population control is arguable. Because of the absolute unpredictability of genius—that might solve all of our problems with one, two, or five discoveries in the next five minutes—it is *never* in anyone's interest to *ever* limit or impede genius. The tribe that is attacking another tribe and hindering their contribution to genius may be so stupid, uneducated, unrealized and self-involved as to not *care* that they are hurting genius, but their stupidity does not change the perfect offer in *not ever* retarding genius in any way. An offer that would be made real if they only expanded their concept of what "tribe" means.

The other interesting aspect about evil and psychopathy is that evil leaders and elites always want the same four things: Power, money, sex and ego (the belief that they are entitled to diminish our lives.) We all want more power, sex, money and ego—but we are sane enough to agree to try to acquire these things by holistically raising our brother and sister's quality of life without breaking the holistic envelope around genius. You can *push* the envelope, but if you break it, that's a crime. Once we see how ridiculous the motivations are of the corrupt, whatever made us makes it that much easier for us to say no to their agenda that is always the diminishment of someone else or our potential.

Each act of genius revealing something new to us about the universe and its components truly is as Dyson said—our genius giving us another insight into something wonderful or convenient about how our universe is wired. Fire, the wheel, ink, paper, writing, electricity, the combustion engine, rubber, oil heat, hot water showers—we are discovering a new way to hot wire something that already existed. Like 2+2 = 4. We didn't *make* that. It already existed, waiting for us to find it. This new "trick" makes our lives more pleasant and helps build a scaffolding of understanding and makes our life that much more linear and predictable. Every layer of understanding helps remove a layer of doubt, stress or uncertainty. The system may always hold back surprises as if to keep us permanently interested and fascinated, as if that is a good thing. Any other system would either have no surprises, or spoil the surprises, or the surprises would fall on the psyche of an animal that really didn't care about how spectacularly each detail lines up in creation to help us.

We have primitive individuals and tribes that do not understand, and want to attack and dismantle, the human instinct to improve its condition and improve upon mistakes. We have hardcore Socialist countries that come to a near standstill creatively and economically except for tools that they use for propaganda to try to convince themselves and the rest of us that some degree of slavery and serfdom imposed by the elite is the best system on earth. Their governments and thinkers are at war with genius because they do not trust it—without ever seeing that what makes genius genius is its utter unpredictability. Maslow's hierarchy of needs points out how more prosperous societies become ever more interested in helping those who need help because of guilt, because of social unrest, and because of the greater opportunity to help. Our neural nets and reflexes are entirely woven to *eventually* adhere to the lesson of mutual respect implicit in genius design because we stress and take on risk and die the more we split from this truly miraculous sleeve of benefit and stress.

This process is an eventuality because to get our epiphany right away, or to be born with it, or to learn it too easily, would be to make the greatest lesson of existence mean too little, and spoil the thunder clap of realization, and the surprise. To cheapen the lesson would also heighten the odds of fumbling away this stunning fact on the interconnection of all things. We would then all the more keep having to relearn the lesson by repeating the same cruel cycle of bad consequences in biological determinism and tribal law. Biological determinism today has a bad rap in science and academics because self-determinism is so much more fun to dream about and play with. Self-determinism is popular because it argues that we can make up the rules, and nothing is sacred, and that there is no God. Self-determinism is popular with the Left because they use it to convince themselves that *they* are entitled to rule and dominate whomever inconveniences them because only they can cure the world by making things easier for those in need. But the world's only problem is splitting from the ideal. Nothing gets easier for anyone who splits from the ideal. The less you help thriving and genius the more you pay a price. *We cannot stop this.* There is no faking—only lag times. Or else we would face no serious consequences for anything we did. You cannot do anything you want to yourself or others and get thriving out of it. Thriving arises only from behaviors and thoughts that holistically help others—for which you are fairly paid, because there is no free ride. The stupefying elegance of this humanity may take sixty thousand more years of "civilization" to see, but our slow pace and stupidity does not change how the deal already lies in the grid beneath our feet.

This invitation to holistically respect and understand everything and everyone that is here is only actualized best through epiphany, education, learning, free choice and the open mind rather than force, violence, disinformation and anti-learning. The lesson is so exquisitely locked in and fair that it might appear to be the sole reason that existence and creation were made in the first place. An engine shows extremely tight parameters of bounded instability because of the nature of metals. The human machine much less so because it is made of flesh and consciousness, perhaps the softest stuff in the universe. But the human organism is a machine no less that shows hardened, bounded instability around this function of thriving for genius and genuis for thriving.

Article 8:
The Center of Human Existence is to Fully Participate in the Genius Machine™

Genius is the Halley's Comet that leads the human race around. Any creature anywhere in the universe would seem to have the potential in evolution to reach this same conclusion. Indeed, what would be the real fun, the real purpose, of any stream of living things that did *not* have this perfect reason for existence and positive unpredictability as their potential? We may understand or rationalize why animals attack the genius machine, but we now see how easily we accept this self-restraint, this self-discipline, in our own tribe in order to be innocent, and to coax the best from one another. But at this point in our awakening led by western civilization two tribes in particular will not extend this universal framework of morality to other tribes. To do so is to harm your own amusement park ride in the genius machine. Our highest success and thriving come from this mission to purely learn, enjoy, study and to create as if in imitation of that which created us. To fail brings on the stress that is psychological, chemical, structural, functional (intra-tribal) or global (inter-tribal).

Until that day comes—voluntarily, enthusiastically—the tuning, the pitch, of the human musical instrument falls that much short of what it could be, and we will experience disruption *by that much*. There has never been a well-done scientific study that does not confirm this holistic design. I say this must be true because how could it be otherwise? Scientists cannot understand why life *evolves*. Life evolves only because existence is only fun and meaningful if it moves toward this complete understanding of unity and purpose, this pure enjoyment of what is. All of life evolves—*except* the operating system that manages each iota of this upward climb. Economics is the level of our own interest and enthusiasm, the entirety of human movement, toward this open end, and how much of ourselves we are willing to give to keep it up. That which is created is destined to create. It is the only process of never-ending surprise and ecstasy. We are pre-fixed to this set-point because it is the most interesting thing we can do, and everything about us is part of it. Anything we fear is in the end only an interference to this most rewarding, safe and entertaining process. Not only is the template innate for this perpetual futurity, but also the sensory framework to *recognize* this lone beacon, this peerless event. You and I agree on a fair price to raise one another's quality of life. If one of us feels that his quality of life is *not* sufficiently raised by the exchange then we do not have to do it. Nothing else is fair. This is why Karl Marx tried three times and failed to describe a Utopia beyond our current course and eventually threw Marxism away. We may argue with our course—but the solution cannot be out-done, or out-thought.

Just as with our health, there are only two types of complaint to take against this OS as it works now: One type is "I wish I could jump over the moon," and the other is, "I wish I could hurt myself or others and get all I want, and live a thousand years." These criticisms of the operating system regret our limitations of matter. However, our limitations of matter are our guides. They are essential to learning. Suffering leads to answers. By this model, *all* of us matter—since, at our best, in the genius machine, we are all either geniuses or helpers of genius. Neither can be their best without the other since it is inefficient to run a genius machine if everyone is a genius and there are no helpers to help with some ventures into genius that require a great deal of help. And there must be payment, since there are replacement costs, and appreciation, incentive, homage, interest, fun, and because mortals must work for everything. Without any of this genius slows, genius stops. —Why bother with the risks and costs? Reward guides and encourages genius. Otherwise there is no fun, no interest, no reward.

Part 3: The Most Perfect Experience Possible to Imagine

Article 9: The Goldilocks Zone

Every level of reality—the Goldilocks Zone—forms the platform or stage from which life moves toward the grandest of realizations. And it *still* gets better, as we shall see. And so it is only a matter of time before the human organism realizes that its highest holistic reward and lowest risk lie *only* in this efficiency when we act as one tribe, and we all consent (as we do now in our own tribe) that none of us should cheat another of their chance to live this holistic experience. All tribe members agree to check their individual wants for the sake of tribe survival and improvement—strength in numbers, cooperation over conflict, in order to have the best start at what comes next. This cohesion holds true even for the most dangerous and destructive tribe of all, a tribe worse than Communists and radical Islamists who seek to destroy everything that is not them.—This is the tribe of elites that is always forming at the top of any group to run it.

Even elites subscribe to the self-control of "honor among thieves" whenever they start standing shoulder to shoulder to plunder, deceive and exploit us, the "subhumans" below for more sex, power, money and ego.

Sometimes people beneath us may indeed be subhuman. But this does not change how the best solution for this problem or any other is to lead with example, education, realization, epiphany and encouragement toward the holistic rather than deception, exploitation, war or peddling superficial artificialities to treat the rot oozing from inside out. The only time war is acceptable and wise is when education fails and fighting is the only option to stop psychopathy that refuses to admit the beauty of the world. This, because they just don't get, or care about, the genius machine. Or they cannot live with the "annoying success" of others who are just plain better at raising quality of life than we are. We accept this annoying success of others in our own tribe because we must for the sake of getting along. This agreement is basic for peace and advancement. We more less accept our rank in our tribe and we do not lie about our gifts and liabilities unless we are insane, and we are more or less willing to hear criticism from members of our own tribe as to how much we are helping the tribe. If all parties are sane, we *want* to hear how well we are helping others, and we want to help fellow tribesmen find their way. This is all part of coaxing the best out of one another—and the mechanics will never change.

Living things anywhere in the universe seem to have to climb through the same stages of realization to see the moral, psychological, physiological and economic perfect: Strength in numbers, cooperation over conflict, for the optimal platform. The word "holistic" only really applies to the tuning of genius, then. This holistic choice to live as one tribe converts us all into potential geniuses or helpers, and would we rather have ten potential geniuses and helpers or seven billion potential geniuses and helpers? This question of numbers of people is relevant when we have a strong movement among billionaire and intellectual elites who believe we currently have six billion too many people on the planet. They hold this opinion even though birth rates are crashing all over the world except in Africa. Even the total unpredictability of genius and its seeming endless variety appear to prove the ideal. No tribe, not primitives, not even authoritarian Marxist elites, accept deadweight among their own on an open-ended basis. Deadweight puts at risk the survival and optimization of the tribe. Only psycopathy fastracks the growth of addiction and welfare deadweight in any "enemy" host tribe that they have targeted for harvest or re-education. Evil can attack this holistic system and oppress holistic reactions with terror and censorhip, and evildoers may time-out and die before all of their holistic violations on the social, psychological and physiological plane catch up with them. We see this often in the cases of Mao, Stalin and Hitler or thugs on the street. Three points should be made: First, remember, nature always arranges organisms, individuals and systems into "spheres" that are entirely self-integrated. Second, the psychic and functional stress of harming anything positively involved in the revelation will stress you even if you are a psychopath or if your tribe is psychopathic. How do we know this? —Are toxins and black mold still toxic to a psycopath? Yes, they are, depending on genetic sensitivity and previous stress sensitivity. Third, just as individuals come with their own sensitivities and strengths, gifts and liabilities, for *everything*, some psychopaths may not have the same level of susceptibility to guilt that more sensitive people have. But they are *always* more unhappy and psychologically healthy. And while their *psychic* violations of the genius machine may not bother them, psyche is still clearly part of our wellbeing, and if the psychopath is a vegetarian, for example, or does great amounts of Tai Chi, or their spine and posture are good—those benefits still do what they do to offset stress in a monster just like they do in anyone else. And let us not forget all the turmoil all of these Socialist monsters kicked up with their murder and terror states. Millions of people would have liked to wait for them outside their hut, but they could not because of secret police. No less, all of this cheapening of life, property, good and evil disturbs economics. It is known that crack cocaine chews wormholes into the brain. It is also known that chronic anger also chews the same sort of wormholes into our brain. Who is more toxic and angry than the sociopaths who cannot stand the perfect order of the world?

All stresses and benefits, as determined by the ideal, go into the hopper, into the balance, to determine our life's holistic fulfillment.

Whether we like it or not, and whether we know it or not. There is only a little variation for your individual strengths and weaknesses that matter less as we zoom out. This determines the exact length of your lifespan. Lifespan is *always* the sum of stress versus benefits acting on our gifts and liabilities. This cannot change, to be fair. This appears to be true because Balance is a structural force of the universe, because of the law of integration, and because stress and benefit are determined by *something* if not survival, thriving and genius. (What else can it be?) Our rights end where the rights of others begin as defined precisely by what is best for the machine. Push back, unrest and revolution are determined exactly by how much any tribe is denying any others to gain access, and to be rewarded honestly by, their help with these three things—unless a false or artificial meritocracy is created by people who are sociopaths the more they pull society away from genius to their own money, power, sex and ego. But the less you care about other people the more you are a psycopath. This toxic dynamic introduces the same iconic tensions. In order to sustain it, psychopaths must struggle for control. They do so by the same means of control: High taxes, high regulations, disinformation, violating the free innate means of exchange between buyer and seller, and addiction; and replacing the meritocracy of genius with the meritocracy of psychopaths, where psycopaths determine success, not genius. Psycopaths are *at least* addicted to their own ego. Usually, they are addicted to the ease and pleasure of something else—money, power, drugs, sex, rock n' roll, food, entertainment, ease, welfare or ideology, which is a tribe believing itself so much better than others that it denies them full access and reward for their genius, like Leftists and radical Islamists. And so we see the Left always push high taxes, regulations, disinformation, interference with trade, addiction, the meritocracy of sociopath elites, with no result but that their elite run amok after power, money, sex and ego no matter what it costs the rest of us. The more anyone is a Leftist the more they will kill you for trying to raise your quality of life in ways they disagree with; in ways that limit the control of *their* psychotics on top.

We matter because the most beautiful process that could ever be conceived depends on us to witness it, and enjoy living it for ourselves. It is far more fun for whatever created existence to have us here to see it, live it, appreciate the splendor, than for us *not* to be here, for the art gallery to be empty, and *no* students in the classroom learning to try creation on for themselves. Hence, sex, kids and continuation mean so much to us. The more any of us *see* how there is no rational alternative to this event and what is good for the machine, the more our resistance becomes hardened against any injustice against this flow, this *Tao,* that culminates in the genius machine. Notice how, as soon as the peerless event of existence is understood, it becomes far easier to resist interference? There is only one saboteur, psycopathy. And they have only one path to interfere—talk you into feeling superior to others enough to interfere with them. They turn people who are whole, or trying to be whole—and turn them into addicted monsters like they are. Addicts who will ignore every iota of proof that what they do and want is not working, and they do not care—because they, at their worst, rise to the singular mission to destroy the perfect. To destroy the single-point meaning of life, the single-point basis of worth, incentive, price and interest. Stress and benefit are ruthlessly tied linearly because they *have* to be in order for us to learn as quickly as possible how to stop hurting ourselves by failing to realize that this linearity is the basis of everything. Ours is the whole life of the highest efficiency, safety, holism, pleasure, interest and self-reward—unless we start down the self-destructive slippery slope of the *artificially* enhanced pleasures and interest in any type of addiction.

Article 10: Antagonists and Agonists, Good versus Evil, Normalcy and Being Whole in the Ideal versus the Abnormalcy in Psychopathy

Understanding evil, strife, stress, pain, perfect irrationality and ruthless psychopathy is much easier when we think of the inevitable roles of agonists and antagonists. How long does it take one hand to wash itself? Without the antagonist of the other hand or wiping the hand on something? One hand cannot wash itself without an antagonist. Only the antagonist makes the feat possible. Even a *chemical* antagonist. If the ideal shows perfect holism and rationalism then it takes little imagination to see that at its worst the perfect antagonist must have all of the potential to match the ideal with perfect irrationality, ruthless inhumanity, and a murderous disregard for the exquisite humanity creation elaborates. This balance might even be more the case if the ideal is so perfect as to test the credulity of the audience (us), in which case the perfect antagonist becomes absolutely necessary. The antagonist is, wittingly or unwittingly, the perfect enzyme, accelerant, agitant, catalyst, stress and irritant to push us out of mediocrity toward realization. This might be especially so in the face of our disbelief that anything not of human making, or made by chance, could truly be perfect, or approach perfection, like the innate morals that govern our world and conscience. The unyielding pain and horror from the perfect foe that evil is, completely controllable only through our own choice, is the most effective spur to jar us awake out of mediocrity.

Our three biggest problems are that 1) we don't realize the offer that is made that locks in consequences, and 2) even now that we do see the stress mechanism we *still* don't know the basics of health that make life linear and predictable with lag times, and 3) we do not yet realize that our only real enemy is not *people* but rather the psychopathy that is the perfect antagonist to the perfect perfect. We know that psychopathy is our only true and definite enemy because otherwise we have to think and act like people who believe that *those people over there* are the cause of all the world's problems. If we don't want to be like them then we are forced not to blame people for the world's problems. As children of western civilization, we know that anyone, with but a mere choice, can so re-align with the genius meritocracy that they are worth keeping around, if we can sway them. Something *in* them distorts their choices, like a prism. That something is an interference to holistic thought that lets us see the first and second laws of truth: First, do no harm, and second, the truth is so big and powerful that it does not take crime to enforce it; it *defines* crime.

We also deductively know—and because we have seen

it!—that nothing has the power like psychopathy to hypnotize people who follow it. They seem *incapable* of believing they could be wrong. (They say the same about us, but we aim at First, Do No Harm, while they aim at "Kill all the people who are whole first, and *then* start doing no harm)." *They will not let go of their hatred for us.* We believe they are misguided and not whole; we do not hate them, but we will not let them harm us or the genius machine. They *hate* us because they believe the world can only be saved by wiping us off the map, and burying our message of peace and genius. We are not redeemable. We are not worth the cost of saving. That might be true of *anyone* except for the utter unpredictability of genius. We think: *What genius will they do if we save them?* The one syndrome in all the world where we lose any care for this single orientation of fact is addiction of any kind. Only the addict will kill itself (and others) and pay no attention to the facts on the way down that progressively scream every alarm at them as they split from the perfect and move toward death. This is suicide! *How can a sentient being ignore all survival reflexes to kill itself?* Only when it is overwhelmed with stress. Thus, sociopathic societies stress their citizens to panic them and mold them into a club to beat back anyone who tries to stop the elite from wrecking the place for money, power, sex and ego. It must follow that either nothing addicts like psychopathy or that addiction is the perfect embodiment of psychopathy, or both. —And that, for the sake of simplicity, God makes only one enemy we have to worry about, that is more vile, ugly, horrid, hypnotic, blind, cancerous, twisted and super- potent than any other. And misery loves company. Addicts seek strength in numbers and cooperation over conflict just like anyone else, and as long as they don't mess up the genius machine who can argue? More addicts means fewer police raids and less chance of being arrested and sent to rehab to feel guilty and stupid. If you love your addiction enough—you'd rather die than go to rehab. We should have no more questions about why evil and pain exist.

The worse the addict, the more they accept the rot that springs from a fountain inside the more we leave the peerless event. This rot springs from a place from where no drugs or surgery or artificialities can do more than scoop out the rot to give us a little time against the inevitable. Drugs, surgeries and artifices *do not heal.* They mask symptoms. They *can* scoop out some rot and give us more time. But they are no substitute for the right way to think and act. The act of living and thinking wrongly is the *only* awful, terminal disease. The rot is the inevitable reward for the violation and is progressive in how it takes up space from what is perfect until we die. Look at how we changed our entire life experience and stress quotient in the 19th century when we realized that our symbiosis with germs was thrown off by filth in our cities and food. The holistic influence of hygiene and better food alone ridded us of nearly 100% of the deadliness and frequency of the disease killers of the ancient world, not the artifice of vaccines. As if an invisible parent was waiting for us to learn how we were better off without snot on our faces—we completely changed our level of stress and well-being and lifespan not by an artifice, but by cleaning up our lives and our food, which means eating more fresh green vegetables. We self-improved in a basic way and changed our circumstances forever. Are you better off with no snot on your face? Are you better off not eating rotten food? *Says who? Who made the rule?* This is almost an esthetic issue that applies to no other animal but *you*. And you *are* better. Every indicator tells you that you are better. I find it fascinating that, if we want to pick one difference maker in our victory over infectious killers, it is Vitamin C.

This is the only ingredient needed to block the most grotesque disease of the ancient world, scurvy, for the way it turned us into bloated, black jello that smelled to high heaven as our flesh and bones and teeth rotted until we died. Scurvy reappears in prison camps with malnutrition. What is fascinating is that we humans are the only animals that do not produce our own Vitamim C. Therefore, our health is completely dependent on the nutrients we get (or our meat products get) from vegetables. I find this very interesting when we humans have a reflex framed around More For Less, while the universe keeps showing us the benefits of Less Is More. This is especially relevant in the next section on benefits.

Similar to cleanliness, research shows there are five more similar "ideals" or hygienes or tunings that each appear to show all the same revolutionary uplift in our understanding and well-being. This is interesting because human lifespan is now known to be 150 years. If stress is the only cause of disease and problems—are we only living half our lifespan because we are dying in mediocrity (by not improving ourselves) where invisible stresses barrage us, make us crazy, kill us, because we have so little understanding of stress, and even less of benefit? What open-ended benefits do we know of that are so good for us that there is no point of diminishing returns the more we use them? Stress opens the door to germ imbalance by reducing immunity. However much we deviate from the ideal makes us "dead." We die when our stress-benefit sleeve moves the marker until our dead space tips the scales against our alive space. Life is just that ruthless and efficient. Ignorance of the law is no excuse. Like stress, facts and psychopathy, germs are unavoidably here to guide us in our search for how to live best. Without them we could live forever in mediocrity. Mediocrity would make little difference if it was just us here. But it is as if something knows better, knows what we are capable of with not so much effort—and pushes us to figure out how to improve ourselves because of the difference it would make.

Article 11: Five Ideals or Hygienes that Defy Hesiod's 'Everything in Moderation'

The five ideals are the "perfect" ways to 1) eat, 2) exercise mind and body, 3) tune our body structure like a musical instrument, 4) relate to others in our tribe, and 5) relate our tribe to others. Should "cleanliness" be considered one of the absolute ideals when germs are inescapable no matter how "clean" we are, and they are absolutely necessary for digestion and countless cellular functions? These five "hygienes" seem to cover all bases to overcome all stresses in human scale, as we said. Only these five ideals contradict Hesiod's ancient axiom of "all things in moderation," to deliver special power in prevention, health, healing and longevity. Nothing medicine has to offer can say the same. Everything Pharma and medicine offers is stressful, toxic. The more you take the more you die. It is only good for emergency procedures. If we converted all of the money in the world into medical care (this figure ranges from $37.6B US to $80B US—either amount makes no difference), none of that medical care would make any difference if you simply decided not to breathe, or eat, or drink, or if you decided to eat, drink, think or act horribly. The more you split from the ideal, the more nothing can be done for you—and the more you die. Medical artifice will never make up for holistic debt. God *forces* you to think and act correctly to get the most out of your life. Voltaire said it best, "Nature heals, while the physician amuses the patient."

The five basics are: 1) a plant-based diet where we are always safe raising the *percentage* of vegetables, fruits and antioxidants in our diet. 2) Placing our body and mind in meditation, mindfulness and prayer at the center of our lives with an exercise program that we see in research on Tai Chi, and less so in Yoga or seated meditation. 3) Fully realizing and implementing the century's worth of findings in chiropractic, and even more so in the innovations of Chiropractic Biophysics (CBP) to precisely analyze and correct deviations from perfect posture/structural tuning, and actually physically correct them. 4) Our *function* in our society—how well we reduce our animal instincts to raise one another's quality of life in our tribe or society. 5) First, within our own tribe, then later, with all tribes. In each instance, the watchword is *less is more.* Less deviation, less quantity, less cost, less fooling yourself with what is fancy and complex, less reliance upon anyone outside yourself. This is very much the essence of thriving. Then you take all of that thriving and do what you want with it, as much or as little as suits your personal taste for work.

Stress unchecked makes the stress cascade that warps our body, makes our body "grow wrong" and rot from inside out. The only real cure is to move the sleeve of how we think and act. This remarkable ideal is even made more startlingly liberal, beautiful, and dare we even use the word "loving"?, because of hormetic effect. Hormetic effect, or hormesis, is a fascinating caveat in the stress cascade. Hormetic effect appears to be a universal psychosocialbiological property, where if we *miss* the ideal by some degree, the stress we bring on ourselves brings on big improvements in our cells. Hormesis means that a little bit of stress, even a moderate amount of stress and poison—become extremely good for us. Therefore, we are not fragile. It as if whatever made us built us to experiment and try different things and not dread each and every thing we do wrong. (How much better can the deal get?) So stress must first push through hormetic benefit and *then* becomes progressively poisonous. Regretably, with NO idea what the ideal is, with NO idea what stress is, with NO insight into true open-ended super-powerful benefits, and with NO historical insight into how often our elites turn sociopathic and begin to hurt us either by intent or indifference—we are corks tossed on the sea of stress. We curse the dark and die horribly at the halfway point, rather than using the five benefits to *completely* change how improved we are. It might be that we can manage our stress with the five benefits and live until 100 or 150 with age- appropriate functionality every year along the way, where age 150 looks like and acts like 90, and 100 looks and acts like 70, etc. Anything less would actually be unfair and impossible to learn from.

Part 4: The Stress Cicatrix of Reality, a Distortion in the Space-Time Continuum?

Article 12: A New Description of the Stress Cascade and How Deviation from Sigma Ruins Our Lives, Elevates Risk and Stress, and Kills Us

Walter B. Canon, Hans Selye, Watmore and Kohi and many others have well-outlined how stress overwhelms human limitations of matter to create stress, alert response, fatigue, exhaustion, dysfunction, faster aging, mental disorders, addiction, disease and death. Stress is difficult to define. I define it as, "A challenge to the way we are supposed to be." Under a microscope, physiologists and anatomists such as Oakley Smith note how distortions in the human prism first begin in the connective tissue. Trauma causes a whorl, a cicatrix, a pellicle, a clenching, a withdrawal, a pathologically distorted spiral in the smooth grain of the tissue that is no longer normally reactive, whole, elastic and youthfully vibrant. Like a leaf scorched by a flame. Stress is a bullet, a torpedo, that has popped a hole in the sphere of our perfect normality. A hole equal to the stress. From that point whatever can go wrong will go wrong, probably in a cancer or octopus of dysfunction "personalized" for this trauma and the process born from it. Each traumatic event forms its own "explosion" in our being, and no two explosions are ever exactly alike. They cobble together as stress builds. *Everything* bad that we do not want, that is wholly contrary to the perfect, comes from this. *Everything* that happens to us is either stressful, neutral or beneficial. *That makes for a ton of stress!* Traumas, poisonings such as we do not eat right, we suffer big or slow injuries, birth traumas, we get dropped on our head by Uncle Floyd, we take on toxins in vaccines, medications, drugs, smoking—and we are off to the races. In any event, the combined most powerful benefits are probably together so powerful that they can offset all other stresses in human scale combined. How do we know?—because otherwise we are quite fragile, and anything else or less would be unfair and make the point of taking care of ourselves and others less relevant as soon as our first big injury occurs, instead of this focus being the most important thing in the universe.

Benefit clearly works positively too in the opposite direction whenever we lower stress, lose weight, start exercising or meditating. If this relationship between benefit and stress is true, then taking care of ourselves (and others) in the best way becomes the most important thing in the world for our own best life. Do we think the scales of Balance only work toward the negative and not the positive? The benefits together are even best to handle stresses out of human scale—like being hit by a bus, or recovering from addiction, or childhood trauma, or waking up suddenly at age 70 to see how much control we have over our health at any time. It will be very interesting to see how strong chiropractic is in regaining ground even near the end of life. The parameters of life are hard enough to learn without confusing them or blurring them even more by making inexact how the "sleeve" between benefit and stress slides back and forth. If we know all of the benefits and stresses then life *must be* linear and predictable around the axis of the ideal, to be fair. Look how much we change our lives when we commit to changing diet, losing weight, meditating and even dabbling in chiropractic. *No one has ever fully committed to CBP chiropractic to try and perfectly retune their spine to see how far this goes.* In chiropractic school, I was told chiropractic would fix 80% of whatever came into my office, 10% would remain unchanged, and 10% would get worse. That breakout was exactly right over nearly 40 years. Then, with CBP, that provided the first means to *actually* measure misalignment and correct it, those results jumped to what looked like 98%. It appeared the only limits were a) the patient quit care, or b) arthritis from old stress blocked the correction. This suggests the *physical* part of the stress complex may, at least temporarily, until broader causative stress is fixed, hold uncanny power to neutralize endogenous stress. In total, the boycott and smear campaign against chiropractic has prevented science from taking up this study while heavy-dose diet, meditation and exercise have been covered by a billion dollars in funding. This unique boycott should tell us something about what some people know about the power of chiropractic by itself. The powers that be will say this is because chiropractic is quackery. Preliminary findings so far say this is absurd. More in-depth study suggests that chiropractic will change everything.

Back to our new stress model: Think of perfect normal and perfect abnormal as two balloons trying to occupy the same space. Perfect abnormal does not even exist until stress begins to introduce it. If the dose of stress is big enough, and the target system vulnerable enough, then the abnormal balloon begins to challenge the perfect balloon. But the perfect balloon has elastic recall, elastic integrity, bend but not break. It is part of a system that is doggedly determined to remain in place with all of its resources to experience the miracle of existence. But it has limitations of matter. If stress is big or prolonged enough (like slumping posture) the abnormal balloon offsets the perfect balloon, or bursts it. At that point, the whorl, the scream of tissues when they are hurt, happens. The proportionate cicatrix or whorl grows in proportion to the ongoing trade- off between stress and benefit. Once the disturbance starts pushing past the hormetic buffer, the influence is entirely bad—or maybe an ongoing hormetic effect continues, but it is not enough to offset the bad. This cicatrix and whorl throws off *everything* about us—it spans our entire spectrum of function, (it must because of the law of integration) chemical, mental, functional, *structural.* And so, it is impossible to holistically address every dimension of this stress-disruption if we allow ourselves to be talked into ignoring how structure and chiropractic are as essential as any other approach to healing. Multi-modal cause, multi-modal fix. This early physical sign of stress is shown in the following illustrations:

Illustrations 1 & 2. Cicatrix, whorl. The cicatrix, whorl or pathological twisting that perhaps marks the first physical signs of the body's normal perfect system "punctured" or "offset" by stress. Reprinted from the Naprapathy Illustrated by Oakley Smith, 1915. Microscope Photo A shows smooth contiguous connective tissue in spinal ligaments. Photo B shows the "kink" in tissue, perhaps a retraction of wounded connective tissue that Smith described as the beginning of structural misalignment in spinal structure.

Stress itself is an affirmation of the stress-benefit model. Stress itself implies better and worse, and if there is better and worse that implies there may be an ideal. And Harrison has pointed out, if there is an ideal there is only one. And its results will stand alone. Chiropractic is the least obvious, most under-appreciated, under-utilized strategy among the five benefits. It is this way because, as we are finally learning now after Covid and the crash of American lifespan aligning with the explosion in healthcare costs and the vaccine schedule—healthcare science, education and media have been held captive since the early 1900s. If this description of the stress cascade is true—where the stress wound is pan-spectrum, and structure is part of that spectrum—then it is hopeless to address our total wellbeing in a holistic manner without the structural part that chiropractic plays. No single silo of the five benefits will clean up the entire mess. They are mutually dependent. No part can be left behind. Response must be proportionate. Just because you do not know that all the hidden stresses in your life are aging you twice as fast as you should be does not mean you are not aging twice as fast as you should be. Chiropractic's impact on the whorl is like nothing else in the world. That is the only reason it has withstood the slings and arrows of discredit, denial and hiding since its discovery in 1895, including the AMA's covert media, academic and government campaign to kill it starting in the 1950s, a conspiracy that the AMA admitted to, and was found guilty of, in 1988. Very cunning people know the chiropractic preliminary findings and they see it will change everything. One study by AVMed HMO in Florida found that the full use of chiropractic in one of their plans reduced drug uptake by 85%. Another AVMed study found chiropractic made 24 of 25 back surgeries unnecessary. The full inclusion of chiropractic will probably make the final leap to unleash the powerplant of our body's ability to heal and manage prevention for practically anything. It would probably offer the lynchpin between cost-effective social welfare and lower taxes and regulations that is the bane of the Left-Right divide.

Regrettably, in history, the powers that be have interfered far too often in our independence and autonomy. This greed and insanity is agonists versus antagonists. Antagonists will always have a purchase on our psyche until we truly come to believe that the human race is one tribe made of many charming sub-tribes is the only efficient way to run things. Our elite have too often won atrocious levels of profit and control from keeping us ignorant. Chiropractic itself shows anti-chiropractic behavior that looks exactly like the behavior from other Pharma-captured institutions. Some even say chiropractic is useless for back pain let alone unable to boost immune levels faster and more sharply than any force ever seen. The atrocity profits taken by some of these elites are fueled by our soul- crushing fear of disease and pain. They grow so astronomically rich from the fear and ignorance they sow that they can keep on pushing the levers to keep civilization in the dark on how simple causes and cures are. All of us have already seen enough to sense that basics in wellbeing are powerful enough that someone is growing rich while making us sicker in order to sell us more stuff that makes us sicker still. They push the artificial, and kill the holistic. Our expectations on how we are supposed to live and think are so perfect themselves that they cannot be tampered with forever. Artifice is complex, expensive, risky, materialistic and never gives you what you want. American lifespan has been falling off a cliff since 2014. It is nowhere near 150 years. Tragically, it is worse since Covid and the atrocity fiasco where *every* recommendation we received for the virus was *exactly* the wrong thing to do to boost our own immunity. Is the stress-whorl in the spines of old people why they shrivel up and fold in half so that they can barely limp along? Do you still think health is still all genetics and bad luck? No real rhyme or reason for anything? We just have to suck up all the suffering because God is that cruel? So enjoy your youth and die young to leave a beautiful corpse? Or do you think that you can believe Pharma and take handfuls of drugs (that are all stress and poison) to stop aging twice and thrice as fast from happening? Or that you can cut away the rotten parts and be as good as new? Few things are as stressful as an operation even if they are very clean: Anesthesia, the trauma of being cut open, the invasion, accidents, the sewing, the healing. Anything but the five benefits are stressful, even air, water and nutrients to an extent, and certainly when you do too much of them. Anything that is not supposed to be in you is an endogenous stress—ten times more toxic than any outside, exogenous, stress. Drugs, Botox and scalpels left behind are *not* supposed to be inside you. They *all* cause inflammation that stresses you and ages you faster. You cannot fool Mother Nature. You cannot fool the operating system requiring utmost holistic respect for all. Nothing is what it is without the totality of what it is, and what it requires, to be here at its best to see the miracle.

The research and the results of chiropractic care robustly suggest that chiropractors and other adherents of structural biology are exactly right in insisting that an ideal of structure, posture, alignment and physical tuning is a key to thriving, and to the care and prevention of anything—even that which is emotional and psychological. The research supports their insistence wholeheartedly. Stress-damage should be thought of as a *physical* warping of our body's prism, of its radiation, since we already know that even light and sound are part of the amazing internal language of our body. Would even the slightest distortion of the prism distort light, sound and radiation? Tuning is that exact. Even slightly off, it is detectable and changing things. No "healthcare system" or society that failed to grasp the delicate and exact interplay of tuning would ever make sense of human suffering and cure. This would also change our concept of helping people since, ultimately, we must all grow our own garden; we must all make our own efforts to respect the ideal. No one can do anything worthwhile for us. I cannot go into the gym and lift my patient's weights for them. I cannot go into their kitchen and eat their spinach. I cannot go into their place of work and raise their brother and sister's quality of life for them and their employer. In the doing of the magic lies the magic. Welfare is a fool's game designed to addict people into voting for elites who have lost all regard for the holistic envelope.

Diet, exercise, tuning, function in tribe, function from tribe to tribe. Structure is probably the last straw that breaks the camel's back because it is the most permanent. Or maybe because that part of consciousness that constructs what is physical is the "hardest," or last, to move? It is much easier to change our mind than move a couch. As soon as our mind changes our entire being changes—but the sofa and our body still have to be moved. It is well established how problematic misaligned joints are. The faster aging part of stress makes the joint age two to three times faster than normal. Autopsy research by Winsor finds that the cause of death is *always* downstream from this break. [1] The bent joint probably represents a comprehensive snarl across the entire spectrum of the individual at that locality that doubtless makes the entire person that much less. Misalignment, to break the camel's back away from flawless structural tuning—like dislocating your finger—might be the best, most reliable red flag indicating a whole scramble of our local normal function. It may be that any misalignment, no matter how slight, will within two weeks (this is how early micro arthritis begins to appear)[2] then ages up to three times faster than our other less-affected parts. The biggest cost for all national healthcare systems is HRPs, High Resource Patients. HRPs, in all nations, is not cancer, heart disease or diabetes, but dysfunction and pain from degenerative arthritis 96% of which comes from structural distortions that begin almost completely in the spine and feet.[3] The other 4% are caused by malnutrition from toxic junk food and too little vegetables. Missing the chiropractic piece means that any combination of other fixes will never wholly work. The stress cascade is like a monstrous hand twisting and ruining every wire in the junction box, starting with the nervous system, the Master System. Ignoring or denying the structural piece dooms us to never reach a holistic model. This, apparently, is just what numerous players in the powers that be want. The proof for chiropractic is extensive, but for length is only touched on here since all other arguments on the human genius machine are innately rational.[4][5][6]

The silo of psychology—knowing and accepting the true meaning of life—is as essential as any other piece of us for alignment with the miracle. If you do not know what the ideal is—how are you supposed to know *exactly* if you are straying from it? All your life you sensed it instinctively— but it does make a difference to know exactly that all of science, psychology, history and economics tracks with the biological determinism that no one here is to be the plaything of someone else. Balance and integration would suggest the totality of our person and social being respond even in radiation and prismatic effect when forces collide. We know this is at least somewhat true unless we believe that the totality of our being is a matter only of chemistry, like Pharma says. If we monkeys could ever manipulate the miracle—could we ever make slavery and serfdom, propaganda and disinformation fair and right? —Or would our efforts turn us into soup instead? What purpose is served by censorship, force, propaganda and disinformation? Who uses them and why? The miracle never needs them.

Part 5: Stress, Benefit and Thriving take their Place Inside String Theory

Article 13:
Biological Determism Anchors Into Thriving and Genius

Some may try to argue that a measured measurable world of biological determinism tied into thriving does not exist in the quantum world and therefore is an invalid argument. Whatever goes on in the quantum world may be a world on roller skates, but no matter how zany that world is at some point it refracts back to Newtonian physics in biological determinism so that we have never seen an exception to the rule that it is a bad idea to jump in front of oncoming buses, and no one is here to the be the plaything of someone else. To me the zaniness of the quantum world is God snapping a lock on the Newtonian physics of biological determinism so that even people of below average intelligence can see the rules on the blackboard of exquisite humanitarianism that, in no way, can be an instance of randomness and chance. This precludes any corrupt individuals from trading bribes among themselves and making happy slaves and serfs cheated of their life potential. The more all of us see the exquisite humanitarianism that comes at the end of every right choice, the harder we are to fool. Something made the benchmark humanity, and we cannot change how that humanity is defined.

Making sense of our world is just not possible without understanding stress and benefit. More stress more death. More benefit more life. Holistic solutions in dose-dependent fashion rip holes in the arguments of those who wish to make less of our lives. Some argue that all our health solutions are chemical in nature. But while it is true that everything is made of chemicals it is not true that chemistry solves all problems. Death by falling is a *structural* problem. Death by poor nutrition or toxicity is a chemical problem. Death by PTSD is a psyche problem from the shock of seeing so much that clashes with the humane superrationality of the ideal. We all possess an innate understanding of how life *should* be. This is why every sane person will refuse to be hurt, to be stolen from, to be deceived, to be denied the voluntary exchange between buyer and seller, and refuse to be addicted so long as they see it coming. No one will accept being on the short end of the stick in any cheat of the holistic envelope. This is why the ideal cannot be argued against. If a logical animal cannot outargue the ideal then it must be accepted.

It is said that Nature abhors a vacuum. What this really means is that there is no free ride. As mortals, we *must* work in order to eat. As mortals, we *must* do something. The wisest, the *only* perfect thing to do, is follow the ideal. The nature of the five benefits show us that, at least in this way, we must make the effort to improve ourselves above our baseline level, and not accept the creature we are born unthinkingly into in the world of kill or be killed, dog-eat-dog. Yes, you have free will, you *can* do that—but look at the results. Probably, no part of us is disconnected from the whole. Stray parts that do not line up in consequence make the lesson so much harder to learn. Whatever the extent of the intervening parts, we can see the overall logic of First, Do No (Holistic) Harm. Both logic and instinct show up everywhere in the ideal. The ideal is impossible to pin down without logic. And it is not that hard. It is not *meant* to be too hard. The truth is *meant* to be found. It is meant to be understood by everyone with at least a seventh grade education because that is the age at which we start easily recognizing friend from foe—especially when we know that the offer from existence is flawless, and that our enemy is *always* trying to make us feel superior to others enough that we will break the holistic envelope.

Since Covid, the militancy of American autocrats has rallied around vaccines like never before. Before Covid, many on the Left would have said that they were highly skeptical of Pharma, and maybe even some vaccines. But since Covid, the militancy of the Left has aligned with the Covid vax and with vaccines in general such that they have become yet another obedience test for the Left. Gone is the Left's vaccine skepticism. How did this almost digital switch take place? The answer lies in the unusual military-like top- down control that the Left has always shown. If the Democrat Party says jump the followers do so unquestioningly—unless the order or policy seems not radical enough, i.e., not expanding or protecting welfare and bureaucracy and Democrat control enough. Democrat Party control looks and operates just like Communist Party control except the taxes and regulations are lower, and the state doesn't own everything, but it does seek to *control* everything through the carrot and stick of outright terror if you stray and bribes taken from the American taxpayer. The other operating motivation for Socialists appears to be the Cloward-Piven strategy: Cloward and Piven were two Communist professors in the 1960s who said that another good way to crash America and its place as the world's only doorstop against Communism was to use unchecked growth of giveaways and welfare and endless immigration to grow welfare addiction and national debt until the budget exploded and the nation was destroyed. Because the Democrat Party is Leftist enough to despise the old holistic America of western civilization that is open-minded and curious about all points of view relative to the meaning of life, it wants to kill old America and replace it with a Socialist one-party, Democrat uni-party model. *That* idea of America is what they are patriotic to.

The Left despises anything that wants it to change. This blind loyalty the Left has always shown is to blame for the body count of 145M people that the Left has killed since 1917. Left is here defined by those who believe that welfare and wealth redistribution and party-run social services (free stuff) are always the bedrock of solving all human ills, and who believe outright terror, even murder, should be aimed at any enemies, which they define as anyone who tries to stop them from saving the world. Communists have *always* called their enemies fascists, even other Communists. If you hear the word "fascist" there's a hardcore Leftist in the room. The West is defined here by people who are open to all ideas about what the meaning of life is because they are uncertain about anything enough that they do believe in the first and second laws of truth, i.e., it probably says First, Do No Harm, and secondly it is probably big and obvious enough that its spread does not require violence of any kind. This contrast between the two sides is backed up by the fact that the West opened the first museums, the Ashmolean Museum in 1683, and the Capitoline Museum in 1734, and the British Museum in 1753. One does not open a museum unless they value the perspectives of others in finding truth. The Left "politicizes" everything. Even their museums. To "politicize" means to propagandize and to layout the punishment state and obedience proofs centered around Party authority, however that might switch at a moment's notice, and no matter what that costs the rest of us. The Left shows a remarkable capacity to ignore all data, history and hypocrisy in their willingness to destroy the opposition, and to go on believing their elites, seen and unseeen. Hence, the Left, it appears, dropped all vax skepticism when their leaders began showing affection for the billions of dollars in bribes that Pharma and Big Food have paid out to grease the skids for articifices that have ruined our health and made us by far the top consumer of medical technology the world over. Pharma raked in $75B in Covid profits. They paid out $1B around Washington. It appears anyone who questions the wisdom and humanity of this is made the target and bombed out of existence. It appears anyone who exposes the maze of corruption, conflicted interests and self-dealing in Washington likewise becomes a media and government agency target to bomb out of existence. In the Covid fiasco, *every* recommendation by our government, medicine and Pharma was *exactly* the wrong thing to do unless you wanted to lower herd immunity and make the pandemic worse. Skepticism for captured leaders, agencies and social media has rightly broken out like wildfire so that anti-Pharma and anti-vax information has become much more easily found and accepted. This terror apparatus has now moved in to sing the praises of vaccines and repeat the mantra that they are solely responsible for turning the corner against the ancient killers and cripplers. But this is not true. It was improved conditions and diet that conquered disease. The impact of more vegetables, fruits and antioxidants in our diet so raised our immune power that 98% to 99% of the lethality and contagiousness of *every* infectious disease happened *before* its vaccine was invented. As if this wretched truth were not enough, it has been argued that the damage to our immune systems by vaccines is responsible for the appearance of childhood cancers that virtually *did not exist* before the vaccine era.[7] It has been said that Pharma raising billions to find a cure for childhood luekemia is like the arsonist promising to find out who started the fire. The following graph shows the timeline as the ancient germ killers became less lethal and infectious before Pharma made a vaccine for any of them.

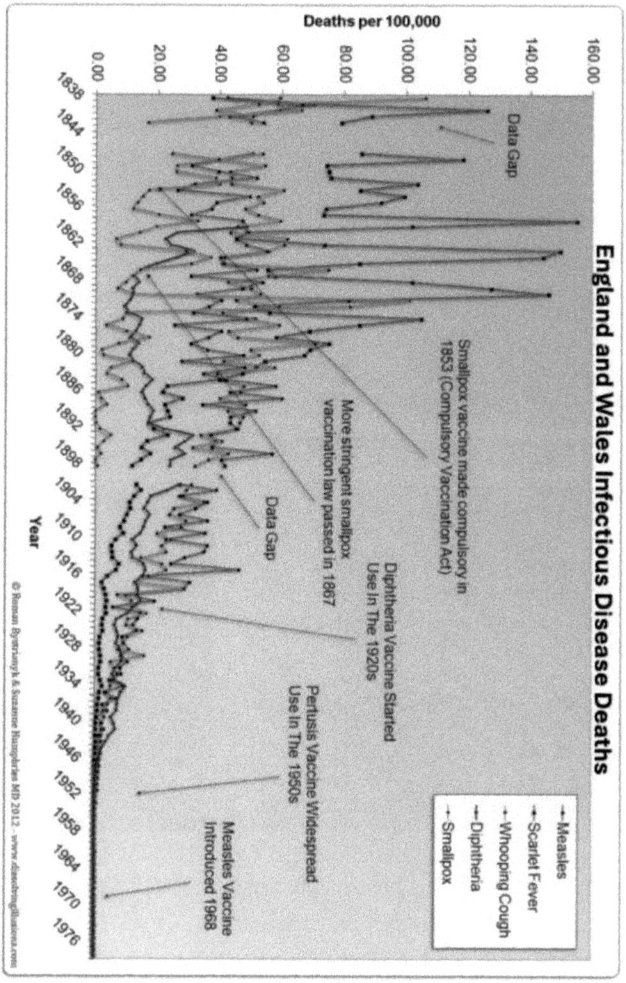

Suzanne Humphries graph on the disappearance of infectious diseases before their corresponding vaccines. Reprinted with permission from Dissolving Illusions: Disease, Vaccines and the Forgotten History, by Suzanne Humphries, copyright 2014, CreateSpace Independent Publishing Platorm.

The following two illustrations are included to indicate the power of chiropractic to change the brain. As Christopher Kent has said, "If we are changing your brain—we are changing you."

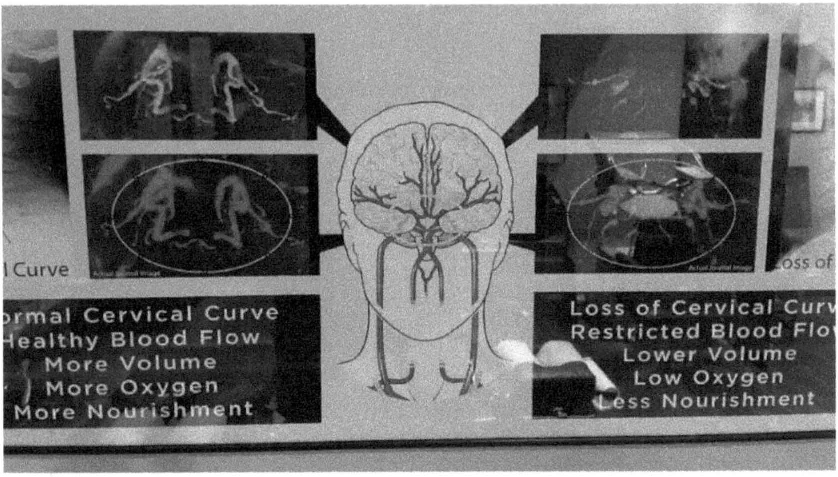

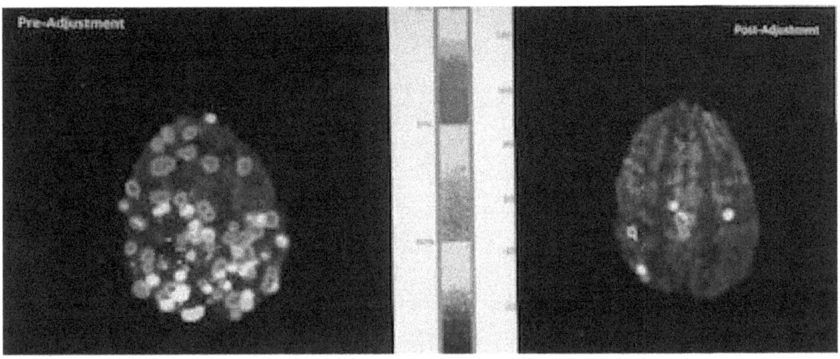

Illustrations 3 & 4: Top images (left-right) pre- and post- fMRI of Circle of Willis vascularity increase from a single Chiropractic Biophysics procedure. This study included three surveys of the medical literature concluding that dementia and Alzheimer's both begin with reduced blood flow to the brain. Bottom brain scan images (pre-chiropractic left side, post-chiropractic right side) demonstrate marked de-stressing of brain activity during stress test after chiropractic adjustment. Case Studies in Chiropractic MRI, Kent C., Vernon L., International Chiropractors Association, 1998.

Article 14: The Operating System is Perfect Except for all the Suffering We Endure In Order to Learn the Perfect System

The symmetry and humane fairness of the operating system is wonderful except for the brutal unfairness of all the suffering and murder we have done to one another in all the thousands of years of man's inhumanity to man, dog-eat-dog and kill or be killed. Think of the tens of millions of people who have suffered from Leftism, for example, from ignorance and psychopathy. It is astounding how much time we have spent being duped by psychopathy and sociopaths. Yes, our horrible suffering guides us—but that does not make the suffering any less inhumane, ugly, painful and unloving. But maybe even this glaring, nightmarish hellscape of suffering and deceit gives us insight into the larger mechanism at work here. Given this horror, either life and existence are *not* perfectly fair and decent, or we are missing something that logic and instinct suggest might be here if the universe is superrationally humane everywhere else except for accident. . . . Let's consider the "in for a penny, in for a pound" argument on the possible exquisite humane circle of the universe if we can explain away this missing chunk.

One possibility is that the offer of existence is *not* perfect, and that brutality and horror are an inescapable, insane and pointless part of existence. But why would it be symmetrically humane everywhere else? Or the other possibility is that it *is* perfect because the awful suffering we experience from accident, ignorance and psychopathy is somehow at some point made up for, apologized for, in some way. It is impossible to argue that the operating system is perfect if there is no such apology and atonement for suffering in, for example, an afterlife of some kind where the horrors make no difference in the long run. Phase One existence is where we are now with tribalism and dog-eat-dog and kill or be killed thinking, and psychopathy progressively ruining everything it touches, and many of us still showing a weakness for wanting to believe it. A possible Phase Two existence would be when we become civilized enough to see ourselves enough as one tribe that we extend all inter-tribe courtesies and acceptance to everyone in all tribes for the optimal performance and surprises from the genius machine. Look at all the cruelty in history resulting from our passage through Phase One where it is impossible for living things to survive without killing something to eat it. Without this afterlife of Phase Three there is no way to argue that the operating system and its offer of the best possible life is superrationally perfect.

I would argue that the few glimpses we have from people who have come back from a look at death support the argument that any suffering we do on this side is made up for on the other side, so that the inarguable unfairness and brutality of whatever happens here during Phase One is compensated for after death in a Phase Three. Otherwise a system that is inarguably perfect in *every other respect* except accident has this glaring hole in it. This is hard to argue because we have never found a means to verify this state that would clinch the argument for perfect humanity. The model is not symmetrical in this one instance while it is gorgeously humane in every other way. Maybe this is where faith in the last unseen part comes in? Nearly all quantum physicists and astrophysicists agree that the large breakthroughs in science point up symmetry, beauty and circularity. Our isotropy being linked to the Genius Machine only closes out in ultra-beauty if Phase Three exists; made perhaps by an entity that shows, by the preponderance of all evidence—no, by *all* evidence—that it super-loves its create-things. The perfection of everything only adds up and turns over full circle if a Phase Three exists for us, waiting to make up for all of the torture we endure for so many mistakes.

In conclusion, we may feel reluctant to accept a functional, processual, experiential and existential centerpoint as the core of why we exist because we can't "see" the girders and beams. But we can see with 20-20 vision how they control our motivations and reflexes. The same structural forces that make a rock a rock, and the air the air, make the centerpoint of our existence just as hard and unrelenting as rock around our desire to be here to witness creation, to appreciate it as much as we can, to learn about as much of it as we are able, and to try creation on for ourselves. Our existence and morality show hardened bounded instability on this and any variants disappear as we zoom out. We are born into this machine. The very existence of fact, instinct, data, knowledge and information prove to be implacably fixed on our own highest humanity that is so divine and well-thought out that the scheme of it has been beyond our ability to conceive for five thousand years of civilization. The entire apparatus is run by a maze of structural forces probably made out of consciousness—that is everywhere and nowhere at once, the substrate of everything, and yet unlikely to ever fully be known, felt or controlled. We can only "control" our place in this scheme to get the most of what is most worthwhile by living and thinking correctly, by fully embracing this model of humanity that stands unchanged apart from us. Only then can we win quantum favor. Complexity science and other branches of science have already pointed out how all of reality is, as Heraclitus said, never placing our foot into the same river. As Zen says, all is impermanence. This is true *except* for the moral lesson behind everything, and that reality sticks our face in. The great lesson that philosophers from Lao Tzu to Marx have missed is the positive role of materialism that both delights us and drives us into lessons on restraint that always end up at the tuning of the Genius Machine that needs, loves and feeds on the problems and errors of mortals. Every success rests on a mountain of mistakes. Which mistake can we remove from any pile that did not help find the solution? Biological determinism is tied to this sweet spot always at the top of thriving. If you want it, you must improve yourself, and earn it. God, Nature, Life, Existence, Creation, the Cosmos, the Universe, or Godness made this the hardened core of the fortress of the universe, of entering into existence. We are not *able* to even imagine an existence for mortals better than this one—and yet we still have dream on top—and anything less would spoil surprises and make learning and humane holistic self-sacrifice mean less than everything.

My son and I once discussed bigotry and supremacy because he said he had learned all about them in school. "Good!" I said. "Tell me the one thing wrong with bigotry and supremacy." To his confusion, I said, "There is only one thing wrong with them." When he gave up, I said, "The only thing wrong with bigotry and supremacy is that you never know when you will cheat yourself of something wonderful."

So it is with any interference in the genius machine.

Author's Summary

This description of the focal center of all existence can be summarized in this way: The human race is best off, and most efficient and secure, as a single tribe made of many charming sub-tribes whose differences stimulate genius far more than sameness ever would. This model minimizes risk, resistance and optimizes interest. Our own personal greed and tastes that make us open to hurting or cheating others are already bounded by how our tribe, and all tribes, define innocence and guilt—so that we are already well acquainted with, and well rehearsed in, how to get along with others and what personal sacrifice is required for optimal tribal function. This "envelope" of behavior optimizes the tribe and the genius machine. To go against it, to challenge it, progressively leads to conflict, resistance, risk, damage, stress and sub-optimization. Our resistance coefficient to being denied or frustrated from what is optimal shows curvelinear modular integrity proportionate to our education and understanding of this perfect humanistic design, this flawless operating system, permanently manifest in existence. The resistance can be seen in any group or individual who realizes they are being actually and holistically cheated of their access to, or their fair reward for, helping genius. Every law of nature and physiology leads us to the conclusion that we must work and think to improve ourselves, to try to be better than we are made or born, in order to get the most out of our own life experience. We may not want to limit ourselves to this eternal discipline—the choice is ours. But the advantages and lower risks in fully holistically respecting others and what is here cannot be made any more explicit than they already are by the slant of the entire table of the universe. At the last, we must voluntarily "go this extra mile" in self-control, since God is not going to actually force us, in light of the fact that forcing us, or making the choice any easier than it is, would rob this paradox of its luster as the most stunning and sublime factor of existence.

Endnotes:

1. The Evidence of the Association, in Dissected Cadavers, of Visceral Disease with Vertebral Deformities of the Same Sympathetic Segments, Winsor H, *Medical Times*, November 1921, pgs 1-7,

2. Plasker E, lecture 2005, Mar 23, book *The 100 Year Lifestyle.*

3. What Contributes Most to High Health Care Costs? Health Care Spending in High Resource Patients, J Manag Spec Care Pharm, 2016; Feb22, Pritchard D, Pretilla A., 2) CDC website, Chronic Disease Indicators, Arthritis; 3) Health.gov, Arhtitis, Healthy People 2030;

4. Chiro.org, archive of chiropractic peer-reviewed research, government reports, studies, white papers, literature reviews.

5. Spokes of Chiropractic Progress, National Chiropractic Legal Action Fund, 2003, 1) 50% working people back problems each year, 2) Back most common disability under 45, 3) every day 2.8M chronically or temporarily disabled w/ back, 4) societal cost $80B, 5) back 2nd most common MD visits, 6) 26M to DC yearly, 7) 1967 Wilson, Chair AMA Section on Orthopedic Surgery - med ed useless, prejudices and controversy, "disastrous sequelae" (see below), 8) 1972 Martin R, Oregon Work Comp (see below), 9) 1979 New Zealand Report "astonishing" DCs under MDs in this field, 10) Scott Haldeman, DC, MD, PhD, NZ Report MD needs 1 year to be DC, physio even more, 11) 1980 testimony John McMillan Mennell MD, medical educator and author, "MS system 60% of human body, med residents if asked would say training here is zero to 4 hours", 12) 1987 US District Court Judge Susan Getzendanner, Wilk decision, Even the defendant's own experts (Mr. Lynk, PhD Economist) assumed chiropractors outperformed MDs, 13) Shekelle P, The Appropriateness of Spinal Manipulation for Low Back Pain, RAND Report, 1991, chiropractic appropriate

6. Objective Physiological Changes and Associated Health Benefits of Chiropractic Adjustments in Asymptomatic Subjects: A Review of the Literature, Journal of Vertebral Subluxation Research, Hannon S, April 2004,* rom, strength, proprio, EMG, immune, beta-endorphin, BP, papillary cycle time, HRV, spirometry, agility, balance, body perception, speed reaction, neurocognitive, visual field, IgA, IgG, IgM, CE4, triglyceride levels, improved cardiac, remission duodenal ulcers, regardless of symptoms, 24 vital measures of health.

7. Day Care in Infancy and Risk of Childhood Acute Lymphoblastic Leukemia: Findings from UK Case-Control Study, Greaves M, Alexander F, British Medical Journal, June 2005, this study supports reduced exposure to infection in first months of life increases risk of ALL, substantiates 1988 same findings,

"isolation inadequate immune priming". First 3 months of natural exposure critical for immune dev; social activity of newborn outside family tied to lowered risk of ALL problems; immune isolation bad; published responses to this article included letters from Roger C Parslow, Senior Research Fellow, Pediatric Epidemiology Group, University of Leeds; Richard Lanigan; John M Grange; Bernd Krone, Klaus Kolmel; Petar Ivonovski; L Travis Haws; Michael Innis—measles infection led to broad spectrum immunity, childhood leukemia before 1920 extremely rare, then around 1940 the DPT vaccine intro correlated to the levels we have today. Aggressive push from WHO, CDC, NIH ties into willingness to deny any connection with broad social health adversity.

www.ingramcontent.com/pod-product-compliance
Lightning Source LLC
LaVergne TN
LVHW050923210926
848482LV00012B/518
9798295455483